Logan Valley Mall Fire
Altoona, Pennsylvania

Investigated by: Thomas H. Miller, P.E.

This is Report 085 of the Major Fires Investigation Project conducted by Varley-Campbell and Associates, Inc./TriData Corporation under contract EMW-94-4423 to the United States Fire Administration, Federal Emergency Management Agency.

FEMA

Department of Homeland Security
United States Fire Administration
National Fire Data Center

U.S. Fire Administration Fire Investigations Program

The U.S. Fire Administration develops reports on selected major fires throughout the country. The fires usually involve multiple deaths or a large loss of property. But the primary criterion for deciding to do a report is whether it will result in significant "lessons learned." In some cases these lessons bring to light new knowledge about fire--the effect of building construction or contents, human behavior in fire, etc. In other cases, the lessons are not new but are serious enough to highlight once again, with yet another fire tragedy report. In some cases, special reports are developed to discuss events, drills, or new technologies which are of interest to the fire service.

The reports are sent to fire magazines and are distributed at National and Regional fire meetings. The International Association of Fire Chiefs assists the USFA in disseminating the findings throughout the fire service. On a continuing basis the reports are available on request from the USFA; announcements of their availability are published widely in fire journals and newsletters.

This body of work provides detailed information on the nature of the fire problem for policymakers who must decide on allocations of resources between fire and other pressing problems, and within the fire service to improve codes and code enforcement, training, public fire education, building technology, and other related areas.

The Fire Administration, which has no regulatory authority, sends an experienced fire investigator into a community after a major incident only after having conferred with the local fire authorities to insure that the assistance and presence of the USFA would be supportive and would in no way interfere with any review of the incident they are themselves conducting. The intent is not to arrive during the event or even immediately after, but rather after the dust settles, so that a complete and objective review of all the important aspects of the incident can be made. Local authorities review the USFA's report while it is in draft. The USFA investigator or team is available to local authorities should they wish to request technical assistance for their own investigation.

This report and its recommendations were developed by USFA staff and by Varley-Campbell & Associates, Inc. Miami and Chicago, its staff and consultants, who are under contract to assist the Fire Administration in carrying out the Fire Reports Program.

The U.S. Fire Administration greatly appreciates the cooperation received from Chief Thomas Sral, Lakemont Fire Company; Joseph L. Lynch, Police Department Coordinator, Fire Coordinator, and Fire Marshal, Logan Township, Pennsylvania; and Trooper Jame J. Behe, Fire Investigator, Pennsylvania State Police, Hollidaysburg, Pennsylvania.

For additional copies of this report write to the U.S. Fire Administration, 16825 South Seton Avenue, Emmitsburg, Maryland 21727. The report is available on the Administration's Web site at http:// www.usfa.dhs.gov/

U.S. Fire Administration

Mission Statement

As an entity of the Department of Homeland Security, the mission of the USFA is to reduce life and economic losses due to fire and related emergencies, through leadership, advocacy, coordination, and support. We serve the Nation independently, in coordination with other Federal agencies, and in partnership with fire protection and emergency service communities. With a commitment to excellence, we provide public education, training, technology, and data initiatives.

TABLE OF CONTENTS

OVERVIEW 1

SUMMARY OF KEY ISSUES 2

BUILDING HISTORY AND OCCUPANCY 2

BUILDING CONSTRUCTION 3

    Roof Deck 4

    Interior Separations 4

BUILDING FIRE PROTECTION 5

THE FIRE 5

FIRE SPREAD 8

CODES 9

LESSONS LEARNED 9

APPENDIX A: Protection, Damage, and Areas of Mall Stores 13

APPENDIX B: Floor Plans of Mall 18

APPENDIX C: Water Test History 22

APPENDIX D: Units Dispatched 23

APPENDIX E: Photographs 26

Logan Valley Mall Fire
Altoona, Pennsylvania
December 16, 1994

Plank Road and Goods Lane
Logan Township, Blair County
Altoona, Pennsylvania

Local Contact: Thomas Sral, Chief
Lakemont Fire Company
309 Orchard Avenue
Altoona, PA16602-4046

Joseph L. Lynch
Police Dept. Coordinator, Fire Coordinator
And Fire Marshal
Logan Township
800 39th Street
Altoona, PA16602

Trooper Jame J. Behe
Fire Investigator
Pennsylvania State Police
PO Box 403
North Juniata Street
Hollidaysburg, PA16648

OVERVIEW

An early morning fire on December 16, 1994 destroyed approximately 20 percent of the Logan Valley Mall, a Regional shopping complex. An effective attack by 59 fire companies successfully controlled the fire although an additional 40 percent of the complex suffered severe smoke and water damage. The fire, which was reported at 2:29 a.m., completely destroyed 15 stores and 9 sales kiosks. The direct loss is estimated at $50 million, with total economic impact of more than $75 million.

The mall, which is located just outside Altoona, Pennsylvania, was built in several stages. It was partially protected by automatic sprinkler systems and partially by a 17-zone heat detection system. The portion that was destroyed did not have automatic sprinklers but the owners were planning to retrofit them. The first fire report came from a central alarm service indicating a fire alarm from the mall. The type and zone of alarm was not transmitted to the central station but was displayed on annunciator panels located outside the complex.

1

The successful control of this fire can be attributed to the large fire suppression force response and effective pre-incident planning. In the planning process the risk of a large fire in the unsprinklered parts of the complex was recognized; effective tactical approaches were incorporated in the plan. The availability of a strong water supply to support the numerous hose streams and master streams, as well as the automatic sprinkler systems, was also a significant factor.

SUMMARY OF KEY ISSUES

Issues	Comments
Fire Origin	Fire began in or near the rear of the largest store in the non-sprinklered part of the Mall.
Fire Spread	Combustible roof coverings over a corrugated metal deck contributed to rapid fire spread. Replacement of original combustible roof coverings did not eliminate the hazard.
Building Fire Protection	Partial sprinkler protection assisted in limiting fire spread. Heat detectors installed in the store of origin operated but did not prevent a large loss fire.
Pre-Incident Planning	Logan Township fire companies planned for a fire through the fourth alarm level and held multi-company drills. Drills included advancing lines and deluge sets into the covered mall area. Standardized training of firefighters helped companies who had never met or drilled together work effectively to overcome this major fire with minimal injuries. However, a concealed structural hazard, which was not a factor in this incident, was identified only after the fire.
Water Supply	A large fire requires a strong water supply. Public water mains, private mains, and a drafting source were adequate for the large volume of water used.
Large Diameter Hose	Firefighters made use of large diameter hose to supply elevated streams, deluge sets inside the building and fireground companies. This improved water delivery to the fire and reduced the time for doing so.
Multiple Alarms	With smoke showing on arrival, the Incident Commander requested additional alarms in accordance with pre-incident planning. This action allowed fire companies to arrive in time to mount a plan to defend at J. C. Penney and to cutoff the fire in the unsprinklered common mall as it traveled toward Sears. The fire rapidly involved fire companies beyond the scope of the pre-incident planning and this taxed radio communication with the dispatch center on the primary fire frequency. Use of the police radio frequency and later a cellular phone overcame this problem.
Covered Mall Separations	Stores with windows, glass doors, walls or other solid separations to the covered mall had less water and smoke damage than stores with open security grills or bars.

BUILDING HISTORY AND OCCUPANCY

The Logan Valley Mall is located in Logan Township, south of the city of Altoona in a commercial area that stretches north along Plank Road. It is near the intersection of Plank Road and limited access highway Route 220. There were no immediate exposures adjacent to the complex.

The Logan Valley Mall is a combination one-and two-story structure containing slightly more than 800,000 total square feet. The overall exterior dimensions were approximately 1,600 feet (east-west direction) by 450 feet (north-south direction). At the time of the fire, it housed two anchor store tenants with the third anchor store vacant and over 100 additional tenants. For discussion purposes, the complex is divided into the East, Center and West Sections (See Appendix B, Figure 1). The fire started in and caused the most damage to the one story portion of the Center Section.

The original parts of the Mall were constructed as a shopping center around 1962. Various additions and renovations were constructed over the years. The common, covered mall between the original stores was added at a later date along with an addition containing stores around a common mall. The largest retail store tenant after the three anchor stores was the 37,000 square foot G. C. Murphy Store, which sold general merchandise. The Cinema IV, Thrift Drug, Crocodile Alley, and Wall to Wall Sound and Video were the next largest areas within the Mall; they ranged in size from 9,300 to 13,600 square feet. (See Appendix A for the complete list of tenants, area occupied, fire protection and fire damage.)

The parts of the East and Center Sections not protected by automatic sprinklers did have a central station monitored heat detection system. It was this system that activated and sounded the alarm which was relayed to the Logan Township Dispatch Center by ADT. Historically, the heat detection system generated a number of false or unwanted alarms. Companies were familiar with these alarms from the Mall and when the second alarm was requested, the Logan County Dispatch Center rebroadcast the alert to the first-due fire companies.

At the time of the fire, plans were being developed for another addition and a renovation which included plans for protecting the original sections with automatic sprinklers. The G. C. Murphy Store, where the fire started, was one of the stores due to be protected with sprinklers.

At the time of the fire, automatic sprinklers protected the two active anchor stores (Sears at the east and J. C. Penney at the center). A third anchor store (Hess) located at the Mall's far west end, was also protected but had been recently vacated. All of the stores and the covered mall area in the West Section were also protected by automatic sprinklers. Automatic sprinklers were also provided in selected areas in the East Section. Besides Sears, these areas included the food court and other stores occupying an area that had previously been a grocery store. Sprinkler heads on this system adjacent to the stores' entrance to the common mall operated during the fire.

The Logan Valley Mall is owned by Crown American Properties, L. P. with headquarters in Johnstown, Pennsylvania. The owner's property insurance is provided through the Factory Mutual System. Regular loss prevention inspections were provided by Factory Mutual Engineering who had made recommendations for providing automatic sprinklers throughout the complex. The Mall appeared to be well maintained and housekeeping was reportedly not a problem. Due to the time of year, storage and stock rooms were heavily loaded.

BUILDING CONSTRUCTION

The east half of the Logan Valley Mall was a one-story, noncombustible building and the west half is a two-story, noncombustible building. Roof construction was lightweight corrugated metal deck supported by steel bar joists resting on either unprotected steel beams and columns or on load bearing concrete block (typically 8 inch) walls. Depth of steel bar joists varied from 16 to 24 inches. The floor to roof deck heights varied between sections of the complex, and between the covered mall and the tenant stores. The most common height within the tenant stores was approximately 18 feet and the covered mall roof was 6 to 8 feet higher.

Various styles and types of typically noncombustible suspended ceilings were provided below the roof bar joists. Most of the ceilings were either a 2 by 2 or 2 by 4 noncombustible tile in a metal grid. Some suspended ceilings were constructed of gypsum board using both metal and wood supports.

Distances between these suspended ceilings and the roof deck varied from store to store in the range of 4 to 8 feet. Similar ceiling types and distances below bar joists were also employed within the covered mall *space*. *Minimal* combustible construction materials were observed in the interstitial spaces above the ceiling and below the roof deck.

Many of the stock and storage rooms in the individual stores, as well as the common utility and mall operations rooms, did not have ceilings. These spaces opened directly to the bottom of the corrugated steel roof deck, with storage using the full building height. Some stores constructed solid floor mezzanines in this space for improved access to storage. Where this arrangement was found, heat detectors were located both above and below the mezzanine. The walls between storage rooms and retail areas stopped slightly above the suspended ceiling level, allowing the storage areas to communicate directly into the space between the roof deck and the suspended ceiling.

The two-story west half had concrete floors poured over metal floor deck supported by unprotected steel bar joists resting on either unprotected steel beams and columns or on load bearing concrete block walls. In store sales areas, typically, a noncombustible suspended ceiling was provided. Light fixtures, HVAC diffusers, and other penetrations were usually not arranged to provide a fire resistance rated floor-ceiling construction.

Roof Deck

The roof covering over the metal roof deck in the area of the fire and adjacent to it contributed to the spread of the fire and the structural collapse. The Center Section was initially provided with a combustible built-up insulated roof covering which consisted of hot-mopped asphalt applied to the metal deck with a fiberglass-asphalt impregnated vapor barrier set into the hot asphalt. A second mopping of hot asphalt attached the fiberglass insulation board to the top of the vapor barrier. An asphalt based built-up roof with a gravel topping covered the insulation board.

Recent re-roofings involved tearing off the combustible roof materials and replacing them with a Class I roofing system[1] with mechanical fasteners for the insulation board. However, the complete removal of the combustible asphalt from the metal deck is difficult to accomplish.

Therefore, even where a Class I roofing system had been installed, fuel supplies for a fire immediately under the metal deck remained. (See Lessons Learned for more on this type of fire.)

Interior Separations

The partition walls between the tenant stores and between the common mall and individual stores varied in their construction type and ability to resist fire, heat, and smoke spread. These partition walls in the stores remaining to the east included hollow concrete block (HCB), HCB with brick facing, and gypsum board on metal studs. Most walls were sealed to the roof deck above the suspended ceiling to minimize the spread of fire and smoke from store to store. Some of these walls were not adequately sealed to the roof deck or contained "poke through" holes which resulted in some additional damage in some stores.

Due to the destruction and cleanup operations in and adjacent to the area of origin, the store separation wall construction in the Mall's Center Section could not be completely evaluated. The east wall

---

[1]A Class I roofing system is a Factory Mutual System approved assembly with limited potential for contributing to fire spread during an interior fire.

of the G. C. Murphy store was constructed of 8-inch HCB which was continuous from the concrete floor to the bottom of an unprotected steel beam that supported the roof. This wall, combined with fire department support, stopped the fire's spread into the adjacent store. In the Center Section, south of the G. C. Murphy Store, the separation walls between stores that helped stop the fire's eastward travel had been removed by the cleanup operation. The separation walls at the south and west extremes of fire travel in the Center Section were constructed of a combination of HCB and brick. See the analysis of fire spread for additional information about these walls.

The separation between stores and the common covered mall also varied throughout the East and Center Sections. It was not possible to identify all separation construction in the Center Section due to the destruction and cleanup operation. However, the G. C. Murphy store did have an open security grill and glass display windows with minor amounts of gypsum wall board over metal studs facing into the common covered mall. Some of the stores used sliding glass door or swinging doors in place of open security grills on their entrances. The heat and smoke damage in stores with doors was less than in stores that used open grills. Above the mall ceiling, the separations generally consisted of gypsum board attached to wood or metal studs. In most instances, both sides of the studs were covered with gypsum board; a few stores were found with only the mall side covered by gypsum board.

BUILDING FIRE PROTECTION

The Logan Valley Mall's unsprinklered areas were protected in 1969 by an ADT Teletherm B 4205 System originally employing only ADT thermopile type heat detectors. Parts of the original system had this style heat detector replaced with UL listed combination 135 degrees fahrenheit fixed temperature and rate-of-rise detectors in recent years. It was reported that the different types of heat detectors were not mixed on the same zone. The heat detection system was divided into 17 zones but it did not identify the specific heat detector in alarm either at the panel or at the device. The system was maintained, tested, and monitored by a central station fire alarm company. The system had a history of false or unwanted alarm activations. It is not known if the point of origin had thermopile heat detectors or the combination fixed temperature and rate-of-rise detectors.

Partial automatic sprinkler protection was installed and plans were in process to complete this protection for the entire complex. The area of fire origin and the adjacent stores and common covered mall were not protected. Six automatic sprinkler heads were observed to have fused as a result of the fire. The fused automatic sprinkler heads were located in the first branch line at the separation between the sprinkler protected store and the unprotected covered mall area. The sprinkler heads were located below the noncombustible suspended ceiling and did not cover the space between the roof deck and the ceiling. The sprinkler systems were supported by fire department apparatus during the fire and all operated satisfactorily.

Around the mall was an eight-inch looped water main supplying fire hydrants and the various automatic sprinkler systems. The loop is supplied by 12-inch water mains in both Plank Road and Goods Lane with a 24-inch feeder main nearby. Flow test records for the complex indicate that more than 2,500 gmp at 20 psi was available on the water main loop. (See Appendix B, Figure 3 and Appendix C for additional information.) Chief Sral indicated that water was not a problem during the fire although the local water authority is seeking a means to charge someone for the water used.

Mill Run Creek at the edge of the complex was also used as a water supply source. Four engine companies drafted from this source and supplied units on the north side of the fireground.

THE FIRE

The fire alarm activation was received by the central station at 2:29 a.m. on Friday, December 16, 1994. At that time, the Logan Valley Mall was occupied by a single cleaning/maintenance worker who was in the west half of the mall which starts at the J. C. Penney Store. Earlier, he had been working in the east half of the mall where the fire started. Prior to hearing the fire alarm, he indicated that no other signs of fire were observed during his shift. The stores had closed at 10 p.m. that evening with no reports of unusual conditions.

The Logan Township Dispatch Center recorded the alarm time as 2:32 a.m. with the dispatch of four Logan Township fire companies to a general fire alarm. The initial box alarm assignment consisted of four township fire companies with four pumpers, one tanker, a squad, a ladder truck and chief officers. (See Appendix D for the dispatcher's log.)

The Lakemont Fire Company, located approximately one-half mile away, is first-due and has the Logan Valley Mall within its boundaries. Lakemont Chief Tom Sral (radio signature 15 Chief in the log) reported on the scene at 2:37 a.m. He responded directly to the north side of the complex where the fire alarm system zone annunciator panels are located on the exterior wall. This area is very near the rear of the G. C. Murphy store where the fire is believed to have started.

Chief Sral observed smoke at the roof line of the G. C. Murphy store and requested a second alarm plus the City of Altoona's ladder truck. He entered the covered mall area through doors on the north side near the Crocodile Alley store and proceeded south to the east-west covered mall segment which passes in front of the Murphy store. There he observed heavy black smoke pouring through Murphy's security gate into the mall and moving into other stores to the south. No flames were observed. He returned to the exterior and directed the first-in engine company to advance a 2-1/2 inch handline with four firefighters through the same mall entrance. A five-inch supply line was stretched from this engine to a nearby loop hydrant.

The first-in engine company had advanced into the structure a short distance when the suspended ceiling tiles began to drop on them. They then backed out to the entrance. About 10 minutes after their arrival, a section of the roof at the rear of the G. C. Murphy store collapsed, venting the fire. About this time Chief Sral requested an "all call" for assistance through the Logan Township Dispatch Center. First with all township companies, the Blair County companies, and then through surrounding county emergency centers.

Pre-incident planning and drills had been conducted at the Logan Valley Mall through the fourth alarm level. Chief Sral indicated the fire almost immediately exceeded these needs. However, the pre-incident work, which included practice advancing handlines and deluge sets with large diameter hoselines into specific points in the mall, allowed for an aggressive combined exterior and interior attack which effectively cut off the fire at strategic locations. The standardized training firefighters received is credited for the fireground efficiency and safety which all of the fire companies exhibited during this incident. Many firefighters had never met before this fire yet they worked as if they routinely drilled together.

Chief Sral became the overall Incident Commander and was assisted by Logan Township Fire Marshal, Joseph Lynch. An informal but workable incident command system was established. Arriving chief officers of other township fire companies were assigned to geographic sectors such as the food court, front of Cinema IV, J.C. Penney rear, and rear corner.

Early in the fire, communications became a struggle between the scene and the Logan Township Dispatch Center. The primary radio frequency, which is used throughout Blair County for the alerting and dispatch of all the volunteer fire companies, is also the primary communications channel between the Incident Commander and the dispatch center. The traffic on this frequency was nearly continuous between alerting, dispatching, and responding companies. Fortunately, Fire Marshal Lynch was able to reach the Logan Dispatch Center on the police radio frequency in his unit. his contact was later supplemented by a cellular phone provided by the county emergency services office.

Fireground communication was facilitated by the use of a second fireground frequency which is common to Blair County companies and officers' radios. While companies from other counties has different primary frequencies, their multiple channel often contained Blair County's primary frequency and sometimes the fireground frequency. There was not a common regional emergency radio frequency.

Deluge sets with solid stream nozzles, mostly supplied by large diameter hoselines, were advanced into the covered mall as planned. Handlines were also advanced into the stores as planned to supplement and support the physical separation of the common mall area from the stores. Elevated master streams were positioned at the rear (north side of the mall) of the G. C. Murphy store and J. C. Penney store to work on the main body of fire and support the separation walls. Additional elevated streams were positioned at the front (south side of the Mall) of the J. C. Penney store and at the Cinema IV for the same purposes.

Four deluge sets were placed inside the J. C. Penney store, two on each floor, to support the separation wall and the glass covered openings into the covered mall. These deluge sets were not called upon to operate and the glass did not fail. Lines were also connected to the sprinkler Siamese connections at J. C. Penney, Sears, and for the food court and adjacent stores.

At least 8 elevated master streams, 7 deluge sets, and numerous handlines operated during the fire attack. In addition to the hydrants supplied by the Altoona Water Authority, four pumpers of 1,000 gpm or greater were drafting from the Mill Run Creek running on the north and west sides of the mall. These units supplied large diameter hoselines to engine and truck companies operating at the rear of the fire. Reportedly some of the water was recycled as the runoff entered the stream and the drafting apparatus sent it back out to the fire again.

Approximately 45 minutes to 1 hour after arrival, a major roof collapse occurred over the G. C. Murphy store and the adjacent stores to the south and west. The signs of impending collapse were identified shortly before it occurred and firefighters were ordered out of the building. Although not previously agreed upon by SOPs, all apparatus on the scene sounded their air horns after hearing the evacuation command on the radio. Following the collapse, interior operations were carefully resumed at the perimeter of the fire. Companies reported that interior visibility improved following the roof collapse due to improved ventilation.

Eventually responding to the fire were 359 firefighters from 59 companies/departments located in five Pennsylvania counties. Some companies traveled over 30 miles to reach the fire. All but one of the fire companies were volunteer. Two minor firefighter injuries occurred: a cut finger and debris in an eye. Neither required hospital treatment. The complete details on the response produced by the Logan Township Dispatch Center are attached as Appendix E. Although companies started to be released at about 8:30 a.m., about 100 firefighters and emergency personnel reportedly remained on the scene through the following Tuesday. Food and refreshments were generously donated by local businesses, the American Red Cross and the Salvation Army.

FIRE SPREAD

The fire was stopped at the J. C. Penney store on the west, the Cinema IV on the south, and the parking lot to the north. It was stopped at the east wall of the G. C. Murphy store which was a non-loadbearing, 8-inch HCB wall. A partial wall collapse occurred at the north end of this wall in the Mall's utility garage area; however, the store to the east received only smoke and water damage.

The fire and heat spread for a considerable distance in the covered mall area from in front of the G. C. Murphy store east to the food court. The covered mall's roof collapsed for about 60 feet to the east of the G. C. Murphy store. Further to the east, a roof section approximately 80 feet long was left standing but showed signs of burning on the underside of the deck. A kiosk below this section suffered only smoke and water damage. East of the standing section, another part of the roof collapsed in a lean-to fashion for a distance of approximately 60 feet. This collapse was due to the roof support beam on the south side rolling over. A candle kiosk under this collapse was not burned and the visible candles were not melted. One basket of candles was covered with melted tar, likely from the roof. (See Appendix F)

The fire damage and collapse of the covered mall roof was attributed to fire spreading under the metal deck roof. The fuel for this fire was the melted and vaporized asphalt from the adhesive and vapor barrier above the metal deck and under the roof insulation. This flammable vapor leaked into the building through the seams in the metal deck where it ignited. This combustion melted and vaporized more asphalt, creating a self-sustaining fire under the roof deck. The undamaged combustible kiosks at floor level further support this observation and also document the effectiveness of firefighter tactics and operations in this area.

The food court had a long skylight covering approximately 20 percent of the roof area. Panels in the skylight were broken during the fire, which relieved the heat and smoke in the immediate area but also drew heat and smoke from the fire in the G. C. Murphy store down the covered mall into the food court. Visible heat damage to a steel beam and bar joists at the northwest corner of the food court's roof indicates a hot gas flow pattern into the skylight. Other evidence of the flow is the steel roof support beam that rolled over and caused a roof collapse directly north of the food court. Areas directly between roof vents and the fire will usually suffer greater damage than other adjacent spaces.

To the south of the G. C. Murphy store, the fire was stopped at the separation wall between The Limited store and B. Moss, another clothing store. (The construction characteristics of this wall could not be identified because it had been removed by cleanup operations.)Neither store was protected by automatic sprinklers. The contents of the B. Moss store received only smoke and water damage even though the roof to the north and west had collapsed.

The Cinema IV was essentially a separate building at the south end of the fire area even though the structure formed part of a covered corridor with the stores to the north. This single story building had HCB and brick exterior walls and was of noncombustible construction. he common wall with the mall was HCB and contained a single double metal swinging door opening. Fire resistance ratings on the doors were not determined and evidence indicates that they did not receive a significant fire exposure. Damage to the Cinema IV was reported to be from water and smoke.

The J. C. Penney Store's separation from the covered mall consisted of a brick-faced 8-inch HCB wall and three large unprotected glass covered openings. Two of these were for show window openings and the third was for the store entrance which was closed at the time of the fire.Due to aggressive fire department actions, the tempered glass covering these openings did not fail during the fire. While not a rated fire wall due to the three openings, this wall was a significant factor in stopping the westward fire travel. The J. C. Penney side of the wall had automatic sprinklers spaced along the openings and the wall. All head spacings were on an ordinary hazard basis. The fire did not cause any heads to operate. No sprinkler heads protected the mall side of the wall and wall openings.

With one exception, the separations, combined with aggressive firefighter support, stopped the fire and heat spread from the covered mall into the stores. Smoke and water damage was significant even where the heat and fire was halted. The one separation failure observed occurred at the wall between the covered mall and the food court. It is believed that the large skylight, broken out during the fire, drew the heat into this area resulting in failure of a main steel roof support beam. There may have been a similar failure at G. C. Murphy to the covered mall separation but due to the destruction this could not be determined.

CODES

The Logan Valley Mall is outside the limits of the City of Altoona in Logan Township, Blair County. According to Township Fire Marshal Lynch, the BOCA National Fire Prevention Code was adopted several years ago. However, no building code has been adopted in the township. The State of Pennsylvania has a labor code that governs exits from structures. Local authorities believed the complex was generally in compliance with the code.

The Lakemont Fire Company and other Logan Township fire companies have conducted pre-incident planning and drills at Logan Valley Mall. Periodic inspections have been conducted and the owner has generally been cooperative.

The application of the 1994 edition of *The Life Safety Code*® NFPA Standard 101, would have required the entire complex to be protected by automatic sprinklers. In addition, the covered mall area would have likely caused the installation of a smoke control system because of exit travel distance limits. These requirements apply to both new and existing covered shopping malls.

The Logan Valley Mall owner, Crown American Properties, L. P., was in the process of preparing plans to complete the automatic sprinkler protection throughout the complex. Additional improvements, such as smoke control, were not identified during the investigation. A new roof covering that was installed the previous summer was specified to be Class I construction. It was applied after the old roof covering had been removed down to the steel roof deck.

LESSONS LEARNED

1. **Combustible roof coverings attached to a corrugated metal deck roof can contribute to rapid fire spread through otherwise essentially noncombustible spaces.**

    This hazard was first published following the 1953 fire in Livonia, Michigan and has reoccurred several times since. An exposure fire heating the corrugated metal roof deck melts and vaporizes the asphalt material commonly used as both the adhesive and part of the vapor barrier. The insulation board, laid into and on top of these layers, and the layers of felt and additional asphalt on top of the insulation do not allow the vapors to escape upwards. As a result, these flammable vapors are forced through the seams in the metal deck panels into the building. The fire inside the building ignites these vapors which continue the process of generating fuel and fire under the metal deck. The heavy black smoke produced often obscures the flames in such a fire. Large caliber hose streams are needed to cool the metal roof deck and supporting structures. Water must reach the deck; steam does not provide the required cooling. Where ceilings are installed, they need to be removed to provide access for cooling hose streams.

2. **Removing combustible roof covering and replacement with Class I or noncombustible coverings does not completely eliminate the hazard.**

    Parts of the complex had new roof coverings which were specified to meet Factory Mutual requirements for a Class I roof covering. Inspections by FM Engineering report that the roofing contractor was following proper procedure for installing a Class I covering. Mechanical fasteners, used with such roof installations, were found in some metal roof deck panels after the fire. The problem is that asphalt remains attached to the metal deck and in the corrugation valleys even when the old roof covering is torn off. This melts and vaporizes, providing the fuel for a fire under the deck. The age and history of the structure's roofing need to be included in pre-incident planning.

3. **Pre-incident planning and multiple company drills in the structure contributed to cutting off the fire's spread and containment.**

    The Logan Township fire companies were well prepared for a fire within the complex. The planning included drills where portable deluge sets and their supply lines were actually advanced and placed. The officers and firefighters who had participated in these efforts were not only able to carry them out but they also provided the leadership for mutual aid companies who had not drilled inside the complex. In operations of this magnitude, it is not practical or desirable for the Incident Commander to direct individual company operations. Sector commanders have to be relied upon to both undertake their tasks and keep the Incident Commander an adjacent sectors informed of actions, progress and failures. In addition, companies must have knowledge of and confidence in the sector commanders. Pre-incident planning and multiple company drills develop skills and this confidence.

4. **A strong water supply is required to support aggressive attack on a fire of this magnitude.**

    The evaluation, analysis and maintenance of available water supplies is a critical element in successful fire suppression. Large undivided areas containing "ordinary hazard" combustible contents will develop into a large fire if not protected by automatic sprinklers. There are several formulas available for estimating water supply needs in these circumstances. These should be used to develop water supply estimates. Flow tests must be conducted to evaluate the quantity of water available and the condition of the delivery means. Where static or mobile water sources employed, drills must be held and sustained long enough to realistically evaluate delivery rates. The information must be updated periodically to accommodate changes in water demands such as from new construction and industry.

5. **Large diameter hose made possible the movement of the water volumes needed to supply the aggressive fire attack in this incident.**

    Large diameter hose supplied engine companies at the fire, elevated streams on the exterior and portable deluge sets placed inside the covered mall. The large diameter hoselines could be quickly located and placed into service to deliver deluging amounts of water. Their use eliminated the need for multiple hoselines and stretches to deliver at or near engine company pumping capacity.

6. **Planning for large properties should include an analysis of the anticipated radio traffic and available frequencies for fireground operations, dispatching, and direction of incoming units.**

In this incident, the primary radio frequency was used for tone activated dispatching, and for communication between the Township Dispatch Center, responding units and the Incident Commander. During the early stages of the incident, radio traffic on this frequency was nearly continuous. Fortunately, Logan Township Fire Marshal Lynch had the ability to reach the Dispatch Center on the township police frequency. This allowed Incident Command to request additional resources even while the primary fire frequency was in use. A portable cellular phone was also placed into service at the command post to provide another communications path. Logan Township and Blair County companies have a separate fireground frequency which assisted with command and control. Many of the mutual aid companies from other counties had the Logan Township primary fire frequency in their radios.

One problem did arise because of the duration of the fire. Portable hand-held radio batteries became depleted and there were problems obtaining enough spare, charged batteries at the scene. Planning must also anticipate the possible duration of an incident and the resources required to sustain operations.

7. **Automatic fire detection has no effect on fire growth and does not prevent large fires from occurring.**

The area of the complex where the fire is believed to have started was provided with heat detectors. According to officials, the system was installed in compliance with the equipment's Underwriters Laboratories listing and was tested and maintained on a regular basis. The type of heat detection device, a thermopile, is considered to respond relatively quickly to flaming fires. However, once at the flaming stage, fires in large spaces grow geometrically with time until either fuel or air supplies limit the growth rate. Fire growth must be controlled in order to stop a fire from becoming large. Heat detectors, and other types of fire detectors in general, may sound an alarm and summon the fire department, but they do not have any capability to control the fire's growth. The major benefit of automatic sprinklers is their ability to control many fires while still small.

8. **Compartmentation is also important in large occupancies.**

Those stores with solid type (windows, glass doors, walls) separations from the covered mall suffered less damage than stores with open grill separations.

Open grill security separations allowed the free passage of smoke and heat into stores from the common, covered mall. This contributed to smoke and water damage to the stores contents. Additionally, the forcible entry through these separations was more difficult and time consuming than for stores with solid separations. Smoke damage in a retail store is often as costly as direct fire damage. Controlling smoke travel is an important element in limiting damage. Rated fire resistance in the separation is not as important as a solid separation with minimal unprotected penetrations. NFPA101, Life Safety Code, does not require separation between the covered mall and the tenant stores, but does require automatic sprinklers and a "smoke control system" within the covered mall. These systems were not present in this incident.

9. **Pre-incident planning should incorporate visits to buildings during construction, major remodeling and expansion projects. Elements affecting the building's structural stability can be altered or concealed during this time.**

During the survey of the damage to the structure, a previously unidentified, concealed, potential hazard was discovered. Near the front of two stores damaged by the fire was a 50-foot long brick faced HCB wall supported by an unprotected steel beam lintel. The wall was 12 to 14-inches thick and 5 to 6 feet high in the ceiling space. This wall was above the suspended ceiling in these stores. Building additions and renovations can result in unexpected structural changes. Some may not present a hazard during normal building operations but can result in dangerous surprises during fires. Construction visits and inspections combined with pre-incident plans can assist in noting such hazards for the future.

Stores with windows, glass doors, walls or other solid separations to the covered mall had less water and smoke damage than stores with open security grills or bars.

APPENDIX A

Protection, Damage, and Areas of Mall Stores

Logan Valley Mall / Logan Township / Altoona, Pennsylvania

Number	Tenant Name	Area – sq. ft.	Protection*	Damage**
	EAST SECTION			
130	Sears (anchor store)	135,464	S	S
114	Thrift Drug	10,020	H	W
116	Hello Shop	2,500	H	W
118	National Record Mart	2,150	H	W
120	General Nutrition Center (G.N.C.)	1,591	H	W
122	Champs	3,728	H	S
126	Centre Film Lab	1,400	H	S
128	Central Bank	3,000	H	S
400	Vacant	2,499	H	S
401	Drs. Grossman & Howells	1,374	H	S
402	Bavarian Pretzel	360	H	W
406	Coldwell Banker Town & Country Real Estate	363	H	W
410	Flower Shop Lottery	49	N	W
412	The Loving Touch	333	H	W
470	Kranich Jewelers	900	S	W
472	McCarthey's	2,105	S	W
474	Rave	2,043	S	W

The following are kiosk tenants in the covered mall

Number	Tenant Name	Area – sq. ft.	Protection*	Damage**
424	Things Remembered	200	N	W
426	Earring Palace	180	N	D
427	The Golden Chain Gang	192	N	W
428	Piercing Pagoda	192	N	D
507	Vacant	180	N	D

*Protection: S = Automatic Sprinklers; H = Heat Detector; N = Neither
**Damage: D = Destroyed; W = Water and Smoke to Contents; S = Smoke to Contents; M = Minimal Damage

Appendix A (continued)

Number	Tenant Name	Area – sq. ft.	Protection*	Damage**
TK1	AK International	216	N	D
TK2	Hickory Farms	200	N	D
TK3	ICM	100	N	D
TK4	Dage	216	N	D
TK5	Fiore Auto	144	N	W
TK6	Piercing Pagota	70	N	W
TK8	Brackeny Leather	192	N	W
TK9	Krunich Jewelry	192	N	W
TK10	J&M Candle	160	N	D
TK11	World Shop	160	N	D
TK12	Kool Foam	168	N	W
TK13	Kool Deans Food	80	N	W

The following are tenants in the food court:

Number	Tenant Name	Area – sq. ft.	Protection*	Damage**
-	Common Seating and Walkway Area	8,385	S	W
442	Vacant	936	S	W
444	Arthur Treacher's	576	S	W
446	Dino's Pizza	840	S	W
448	Long Island Deli	726	S	W
450	Vacant	782	S	W
454	Skooters	590	S	W
456	Vacant	1,328	S	W
460	Vacant	964	S	W
462	Wong's Wok	734	S	W
464	Dogs Etc.	533	S	W
466	Fries Etc.	539	S	W

	CENTER SECTION			
101	J. C. Penney (Anchor store)	162,993	S	S
102	Payless Shoe Source	4,000	H	D
103	Mall Barber Shop	765	H	D
104	The Finish Line	3,375	H	D
105	Mall Shoe Repair	1,115	H	D

Appendix A (continued)

Number	Tenant Name	Area – sq. ft.	Protection*	Damage**
106	G. C. Murphy	37,000	H	D
107	Vacant	440	H	D
108	Lane Bryant	5,160	H	W
109	Kay Jewelers	1,884	H	D
110	Foot Locker	2,450	H	D
111	Kinney Shoe	4,293	H	D
112	Shoe Department	4,919	H	W
113	Casual Corner	3,680	H	D
115	American Outfitters	3,680	H	D
119	B. Dalton Bookseller	2,453	H	D
121	Limited	4,292	H	D
123	B. Moss (formerly Brooks Fashion)	2,858	H	W
125	Ormond Shop	8,211	H	W
139	Your Hair Connection	1,102	H	D
204	Crocodile Alley	9,306	H	D
600	Cinemas IV	13,634	H	W
606	Convention & Visitor Bureau of Blair County	2,042	H	S
WEST SECTION				
800	Hess's (vacant anchor store)	105,323	S	M
700	Meyer Jonasson	2,846	S	M
702	Card Factory	1,599	S	M
703	Non-lease Storage	627	S	M
704	Lady Footlocker	1,600	S	M
706	The Kitchen Works	1,882	S	M
712	Victoria's Secret	3,979	S	M
710	Freshens Yogurt	748	S	M
714	Lamens Furniture	1,892	S	M
715	Central Bank	2,122	S	M
716	Shirley's Shoes	1,840	S	M

*Protection: S = Automatic Sprinklers; H = Heat Detector; N = Neither
**Damage: D = Destroyed; W = Water and Smoke to Contents; S = Smoke to Contents; M = Minimal Damage

Appendix A (continued)

Number	Tenant Name	Area – sq. ft.	Protection*	Damage**
718	Naturalizer	1,002	S	M
720	Vacant	6,640	S	M
724	Gordon Jewelry	1,050	S	M
726	Where On Earth	1,250	S	M
728	Claire's Boutique	748	S	M
734	Kay Bee Toy & Hobby Store	3,488	S	M
736	The GAP	4,732	S	M
740	Payless Shoe Source	2,977	S	M
742	Pearle Vision Express	2,977	S	M
744	Thom McAn	2,654	S	M
746	Jean Nicole	5,954	S	M
750	Fines	3,572	S	M
731	Kids I.D. & Specialty Shop	540	S	M
732	Gardner Candies Store	1,224	S	M
752	Musselman's	1,775	S	M
753	Non-lease storage	412	S	M
754	Topkapi	748	S	M
758	DEB	7,273	S	M
762	County Seat	3,800	S	M
763	Management Operations	unknown	S	M
The following are second floor tenants:				
900	Panache Hair Studio	1,186	S	M
901	Vacant	1,165	S	M
902	The Photo Factory	1,382	S	M
903	TNT Shirts	742	S	M
904	Vacant	1,173	S	M
906	Pirates Clubhouse Store	1,186	S	M
908	Vacant	1,977	S	M
909	Vacant	365	S	M
910	Hickory Farms	2,012	S	M
914	Vacant	3,891	S	M

Appendix A (continued)

Number	Tenant Name	Area – sq. ft.	Protection*	Damage**
916	Vacant	3,333	S	M
918	Vacant	2,002	S	M
920	Dino's Pizza	1,000	S	M
922	Blair County Resource Center	3,663	S	M
924	Orange Julius	799	S	M
926	Vacant	804	S	M
928	Vacant	1,200	S	M
-	Non-lease storage	2,042	S	M
930	Only $1.00	2,530	S	M
932	Vacant	2,346	S	M
934	Waldenbooks	2,602	S	M
938	Ye Olde Hobby Shop	5,204	S	M
940	Radio Shack	2,394	S	M
942	Pet World	1,977	S	M
944	Matthew's Hallmark	3,227	S	M
945	Wall to Wall Video Storage	584	S	M
946	Spencer Gifts	2,706	S	M
948	Xpressions Unlimited	1,800	S	M
950	The Wall	10,455	S	M
956	Holiday Hair Fashion	1,050	S	M
958	Spectacles	1,050	S	M
The following are kiosks in the covered mall:				
TK14	Mountain Magic	192	S	M
TK15	Keyboard World	240	S	M
TK16	Ross Enterprize	100	S	M
TK17	All American Sweats	192	S	M

In addition to the above, there is approximately 115,600 square feet of covered mall, service corridors, mechanical rooms, public (restroom) facilities and operations area.

*Protection: S = Automatic Sprinklers; H = Heat Detector; N = Neither
**Damage: D = Destroyed; W = Water and Smoke to Contents; S = Smoke to Contents; M = Minimal Damage

APPENDIX B

Floor Plans of Mall

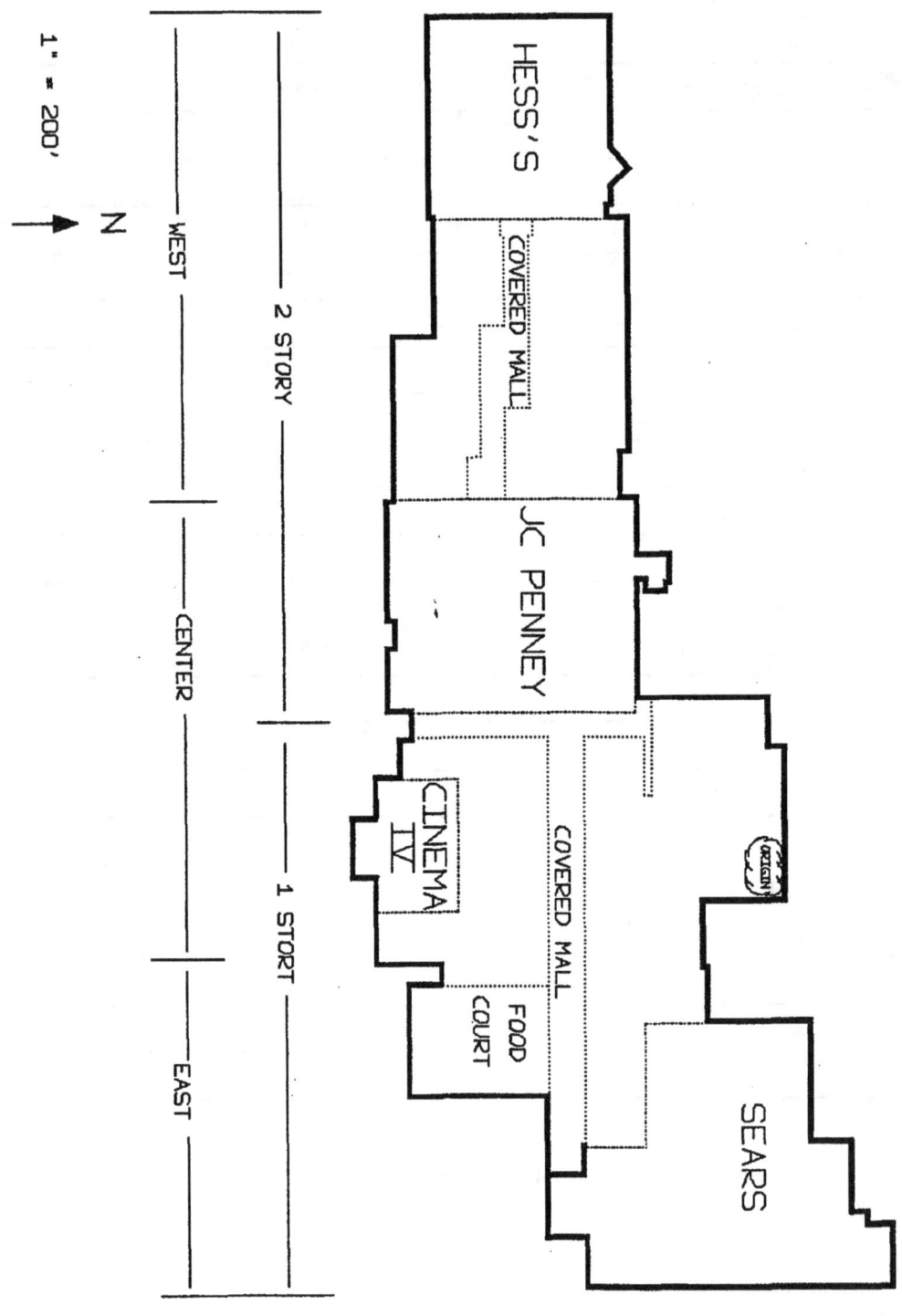

Appendix B (continued)

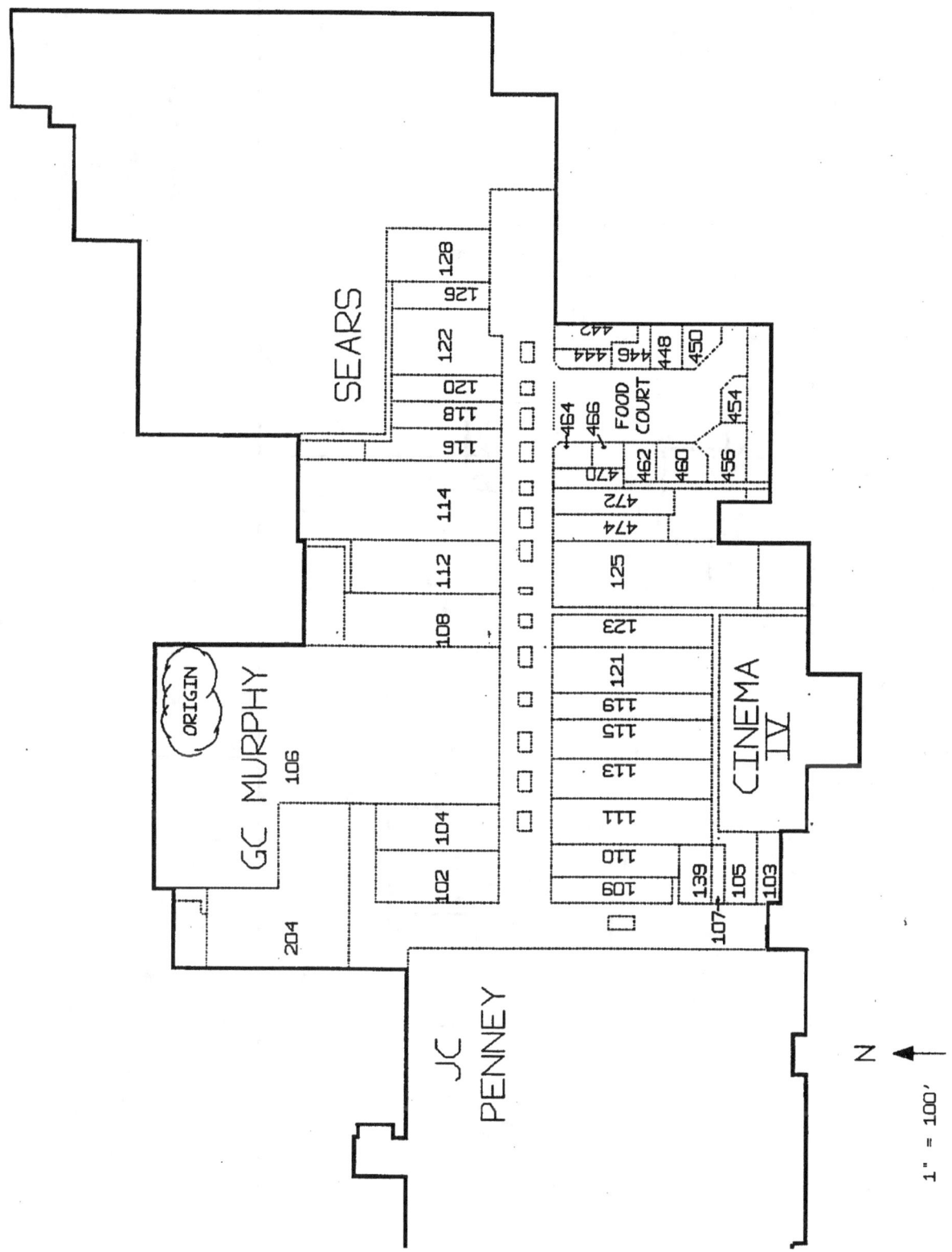

Appendix B (continued)

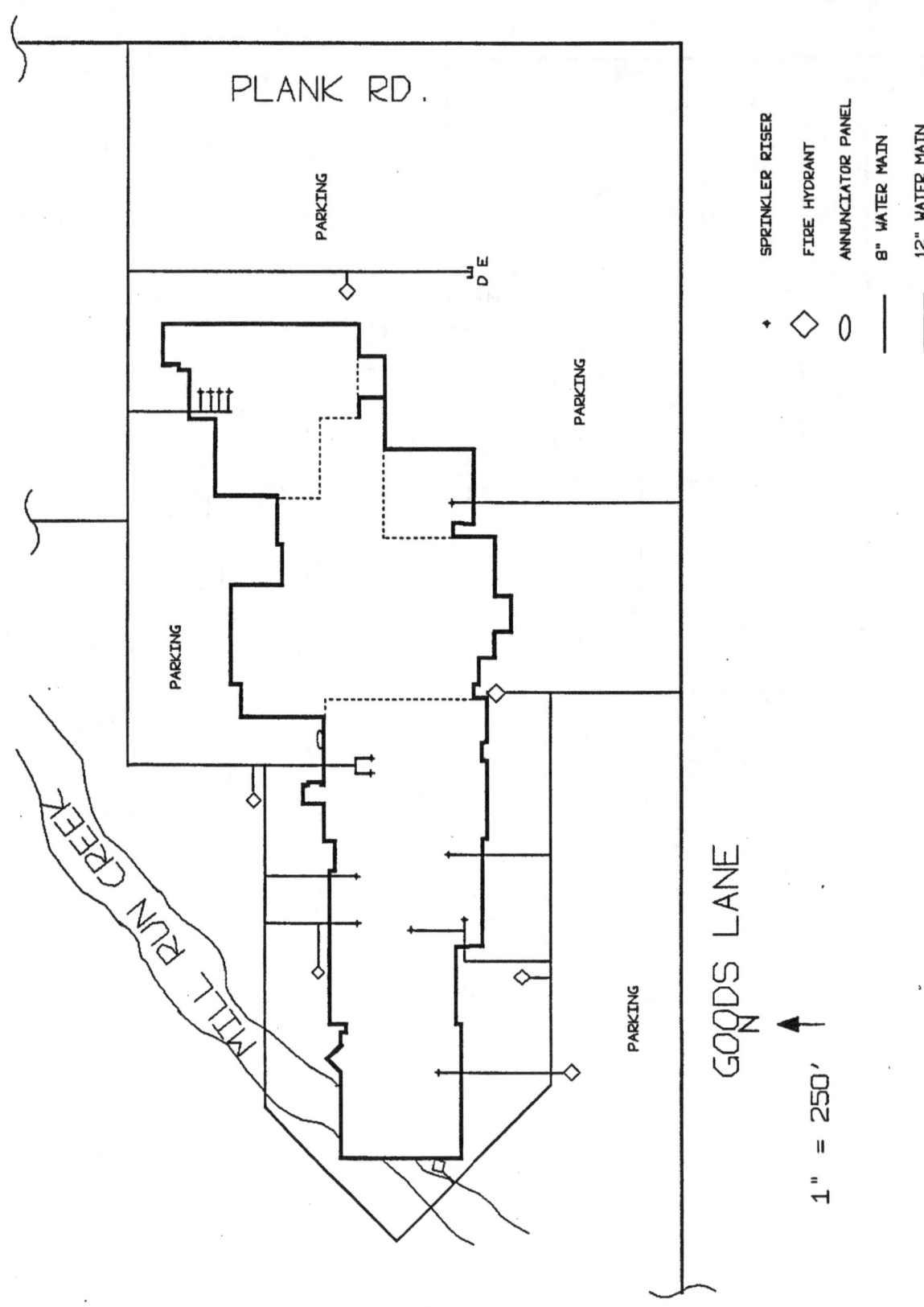

Appendix B (continued)

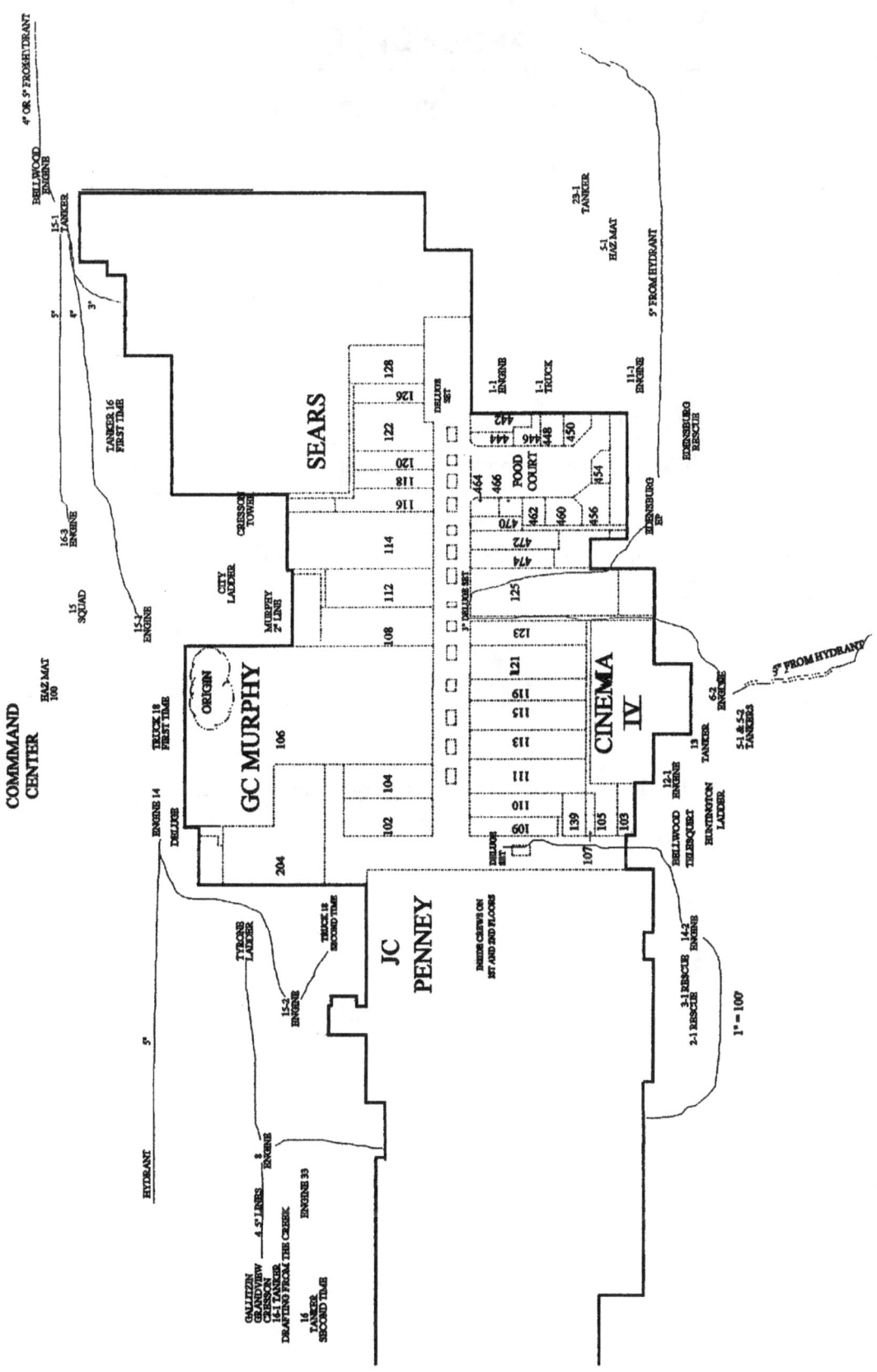

APPENDIX C

Water Test History

Water flow test information:

Factory Mutual Engineering, April 1990 for hydrants at both the east and west end of the complex:

115 psi	Static
195 psi	Residual
0.8	Hydrant discharge coefficient
850 gpm	Water flow with one hydrant outlet

Altoona Water Authority, April 1989 for hydrants at about the same locations above:

120 psi	Static
54 psi	Residual
2,000 gpm	Water flow

No information was available regarding hydrant discharge characteristics.

APPENDIX D

Units Dispatched

Response from Blair County; Logan Township:

Company Number	Company Name	Number of Firefighters
11	Juniata Gap	13
12	Newburg	10
13	Mill Run	10
14	Kittnaing Trail	17
15	Lakemont	17
16	Greenwood	19
17	Grandview	16
18	Aerial Truck #18	2
24	Sinking Valley	11

Total Firefighters: 115

Response from Blair County; Hollidaysburg 9-1-1:

Company Number	Company Name
1	Hollidaysburg
2	Duncansville
3	Geeseytown
4	East Freedom
5	Allegheny Township
6	Roaring Spring
8	Martinsburg
9	Williamsburg
21	Tipton
22	Bellwood
23	Pinecroft
32, 33, 34	Tyrone (Hookies, Neptune, Citizens)

Approximately 100 Firefighters

Appendix D (continued)

Response from Blair County: City of Altoona:

The City of Altoona responded with their 100 foot aerial truck, Chief Santone, and Deputy Chief Weimer with the Hazmat Command/Communications Vehicle.

Blair County Emergency Management Agency was notified, activated the Emergency Operations Center, and responded to the scene.

Response from Bedford County:

Company Number	Company Name	Number of Firefighters
31	Bedford	4
34	Hyman	2
36	Six Mile Run	3
37	Shawnee	4
38	Alum Bank	3
41	Southern Cove	3
42	Imler	4
45	Cumberland Valley	2
55	New Baltimore	1

Total Firefighters: 26

Response from Huntingdon County:

Huntingdon County paged the entire county and had a response from approximately twenty fire departments that included 75 firefighters.

Response from Centre County:

Centre County dispatched the aerial truck from Alpha Fire Department to the Tyrone area for coverage of the northern section of Blair County.

Response from Cambria County:

Tower 707 from Cresson

Snorkel 223 from Dauntless

Utility 225 from Dauntless

5-Engines

4-Rescues

4-Squads

Total Estimated Firefighters: 130

Appendix D (continued)

EMS Response

AMED

Hollidaysburg Ambulance

Duncansville Ambulance

Hollidaysburg 9-1-1

Altoona

PSP-Hollidaysburg

Bedford County 9-1-1

Cambria County 9-1-1

Centre County 9-1-1

Huntingdon County Dispatch

Police Response

Logan Township

Altoona

Pennsylvania State Police-Hollidaysburg

Allegheny Township

Martinsburg Boro

Blair County Sheriff

Bedford County Sheriff

Special Response:

Blair County Fire Police

American Red Cross (Altoona)

AMTRAN

Salvation Army

Blair County A.R.E.S.

Dispatch Centers:

Logan Township

The following is a list of businesses that supplied food, services, materials, etc. throughout the disaster at the Logan Valley Mall.

McDonalds; Plank Road & Station Mall

BILO

Sam's Club

WalMart

Eat & Park

Ultimate Bagel

Taco Bell

Sheetz

Kentucky Fried Chicken – Plank Road

TGI Friday

Dominoes Pizza

Ramada

Roudabush Bakery – Claysburg

Super 8

Red Lobster

Frantello's – Hollidaysburg

Pizza Hut

Mrs. Groves

Al's Doughnuts

APPENDIX E

Photographs

Photo 1. East wall of the J. C. Penney Store where the westerly fire travel was stopped.
Roof line of covered mall is identified by the dark black horizontal line.
Glass entrance and show windows are behind the chipboard
and wood studs.

Appendix E (continued)

Photo 2. Covered mall to the east of the Murphy store and some of the stores adjacent to the
covered mall. The three different roof heights and construction features
are illustrated. The tenant separation wall for the right hand store
has been removed.

Appendix E (continued)

Photo 3. Looking east into the covered mall from about where the forklift is located in Photograph 2. The ceiling tiles were removed during the fire and the floor debris was removed before the photograph was taken. Note the partial roof collapse in the background.

Appendix E (continued)

Photo 4. Store on the south side of the covered mall. Typical of the forcible entry work
performed on the security grill and the damage to goods.
This store had automatic sprinklers and the sprinkler heads
at the front entrance operated.

Appendix E (continued)

Photo 5. Open sprinkler head inside the store in Photograph 4. The ceiling tiles and grid
around the head were removed before the picture. Typical concrete block
tenant separation wall and metal deck roof construction in the older section.
Interior tenant construction has used wood studs for the metal ones.

Appendix E (continued)

Photo 6. Moving further east in the covered mall, the counters in the foreground are
also in Photograph 3.This area is where the metal roof began a partial
collapse. The goods in the kiosks were either stored before
the fire or removed prior to the picture.

Appendix E (continued)

Photo 7. A candle kiosk under the roof collapse area shown in Photograph 6. Note that the candles, wicker baskets and table skirting are not melted or burned. Yet, the metal bar joist is severely deformed.

Appendix E (continued)

Photo 8. A front view of the partial roof collapse over the Things Remembered kiosk.

Appendix E (continued)

Photo 9. The top of the metal deck roof illustrating the insulation board, the vapor barrier
under the board, the seam in the deck and the deformed condition
of the metal. This section is in the partial roof collapse area
shown in previous photographs.

Appendix E (continued)

Photo 10. The underside of the metal deck r of above one of the stores adjacent to the covered
mall. Note the tar dripping through the seam in the valley between metal
deck sections. You can also see three mechanical fasteners poking
through ridges in the deck. Such fasteners are usually found
when less combustible roofing methods are employed.

Appendix E (continued)

Photo 11. Example of the hidden structural condition found above the suspended ceiling in one of the stores adjacent to the covered mall.This is a brick faced hollow concrete block wall supported on a 16-inch wide unprotected steel lintel that spanned the full width of the store.